Copyright Presh Talwa

About The Author

Presh Talwalkar writes math puzzles and articles on game theory. The *Mind Your Decisions* YouTube channel features videos on math. He studied Economics and Mathematics at Stanford University.

Books By Presh Talwalkar

The Joy of Game Theory: An Introduction to Strategic Thinking. Game theory is the study of interactive decision-making, situations where the choice of each person influences the outcome for the group. This book is an innovative approach to game theory that explains strategic games and shows how you can make better decisions by changing the game.

Math Puzzles Volume 1. This book contains 70 interesting brain-teasers and fun math problems in counting, probability, and game theory.

But I only got the soup! This fun book discusses the mathematics of splitting the bill fairly.

40 Paradoxes in Logic, Probability, and Game Theory. Is it ever logically correct to ask "May I disturb you?" How can a football team be ranked 6th or worse in several polls, but end up as 5th overall when the polls are averaged? These are a few of the thought-provoking paradoxes covered in the book.

Table of Contents

Preface...1

Chapter 1: Introductory Examples...2

Chapter 2: The Steps To Multiply By Lines..................................13

    2.1 Single Digit Numbers...13

    2.2 Two Digit Numbers..14

    2.3 Three Digit Numbers...19

    2.4 More Than 3 Digits..25

Chapter 3: Special Cases..27

    3.1 The Digit 0..27

    3.2 Three Digits Times Two Digits..34

    3.3 A Large Digit...37

    3.4 Many Large Digits...41

Chapter 4: Rotating The Diagram...47

Chapter 5: Why Multiplying By Lines Works..............................54

    5.1 Line Intersection Is Multiplication......................................55

    5.2 Bundles Of Lines..55

Chapter 6: Advanced Examples..62

    6.1 Algebraic Expressions..62

    6.2 Complex and Imaginary Numbers......................................66

    6.3 Negative Numbers..70

 6.4 Binary Arithmetic..71

Conclusion..73

Chapter 7: Answers To Exercises..74
 Chapter 1...74
 Chapter 2...79
 Chapter 3...84
 Chapter 4...89
 Chapter 6...90

More from Presh Talwalkar..91

Preface

In May 2014, I made a YouTube video about how to multiply numbers by drawing lines. By the end of the month, the video received over a million views. I was surprised by the response and many people were interested in learning more.

Hoping to find an instructional manual, I researched the origins. The method is commonly referred to as "how the Japanese multiply" or "Chinese stick multiplication." But no book, scholarly paper, or credible source in my research corroborated the method is taught in Japan or China. Furthermore, none of the sources even mentioned the mysterious technique. Adding to the puzzle is that multiplying by lines is not similar to historically used methods for multiplying, such as the abacus or the Egyptian method/Russian peasant multiplication. Multiplying by lines is most similar to lattice multiplication and the grid method of multiplication, but the procedure to draw lines and make the multiplication process visual is a noteworthy innovation. The earliest reference I could find to multiplying by lines—in print or on the web—is a video dated November 2006. But no source I have come across has documented the method's origins or systematically elaborated on its applications.

I wrote this book to illustrate how you can multiply by lines, to enumerate the precise steps in the process, and to offer examples of how to use the method. I also present novel applications of how one diagram can solve additional problems and how multiplying by lines can be used for algebraic expressions. I have included exercises in several sections and provided answers in the last chapter.

Chapter 1: Introductory Examples

How would you multiply 12 by 13? You probably learned a method similar to one depicted in the following figure.

$$\begin{array}{r} 12 \\ \times 13 \\ \hline 36 \\ 12 \\ \hline 156 \end{array}$$

The standard algorithm is taught in American schools, it works every time, and it is worth learning.

This book is about explaining a different, visual method. You can multiply numbers by drawing lines and counting intersections. We will first illustrate the method.

Let's multiply 12 by 13 using lines. The first step is to write the numbers and draw a single slanted line to correspond to the digit "1" in the number 12.

12 x 13

Next, leave a little bit of space and draw 2 lines for the "2" in 12.

12 x 13

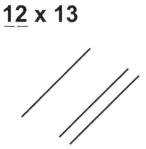

For the number 13, draw lines in the opposite direction. First draw 1 line for the "1" in 13.

12 x 13

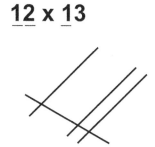

Then leave some space and draw 3 lines for the "3" in 13.

12 x 13

Group some of the intersections together, separating them with lines.

12 x 13

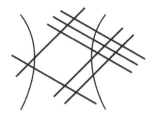

Now count the number of intersections in each group. On the right, there are 6 intersections between the lines.

12 x 13

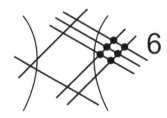

In the middle, there are 5 intersections.

12 x 13

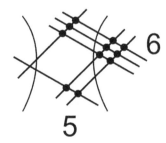

And on the left, there is 1 intersection.

12 x 13

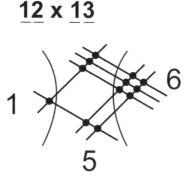

Finally we can determine the answer by writing the number of intersections as a single number, counting from left to right. There is 1 on the left, then 5 in the middle, and 6 on the right. Writing that as a single number, 156, is our answer. That is, 12 × 13 = 156. You can double-check with a calculator; that's the correct answer!

12 x 13 = 156

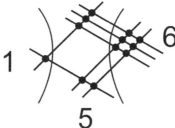

Multiplying by lines works almost like magic and it is so different from what is taught in schools. Many people suddenly get excited about math and want to do more examples.

So let's continue with another problem. Let's do 15 by 21. The process starts out the same.

Start by drawing 1 line for the digit "1" in the number 15.

Then leave some space and draw 5 lines for the "5" in 15.

For the number 21, draw lines in the other direction. Draw 2 lines for the "2" in 21.

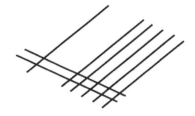

Then draw 1 line for the "1" in 21.

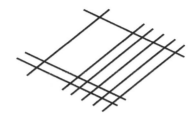

Now group the intersections by separating them with lines.

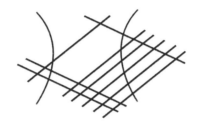

On the right there are 5 intersections, in the middle there are 11, and on the left there are 2.

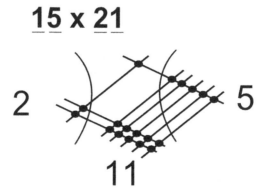

In this problem, we need to make a small adjustment before writing the answer. Because there are 11 intersections in the middle, we need to "carry over" the tens digit "1" to the left. So the 11 becomes a 1, and the 2 in the next set of intersections becomes a 3. (This is similar to the carry over process in the standard multiplication method.)

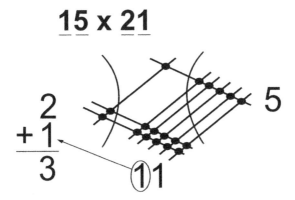

Now we can determine the answer. Write out the number of intersections as a single number, combining the numbers in each group from left to right. There is a 3 on the left, then a 1 in the middle, and a 5 on the right. Writing that as a single number, 315, is the answer for 15 × 21. And once again, that's the correct answer!

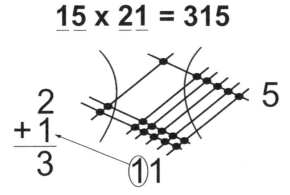

To illustrate the power of multiplying by lines, we will do one more introductory example. Let's say you want to multiply the numbers 123 and 321.

The process starts out the same as the previous examples. To draw the number 123, we draw 1 line for the "1" in 123, then we will leave some space and draw 2 lines for the "2," and finally, we will leave some more space and draw 3 lines for the "3." So the number 123 can be drawn as follows.

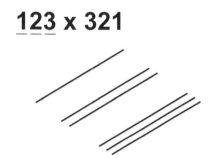

For the number 321, we need to draw lines in the opposite direction. We draw 3 lines for the "3" in 321, then we will leave some space and draw 2 lines for the "2," and finally, we will leave some more space and draw 1 line for the "1." So the number 321 is drawn and our diagram looks like the following.

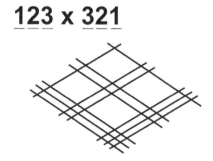

Now we need to group the intersections that are vertically aligned. In this problem, there are five groups of intersections that are vertically aligned.

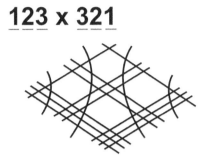

We can count the number of intersections in each group. Starting from the far right, the number of intersections in each group is 3, 8, 14, 8, and 3.

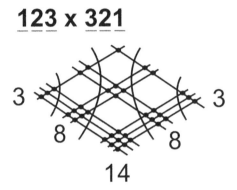

We again have to make a small adjustment before writing the answer. We need to carry over the tens digit "1" in the number 14. So the number 14 becomes 4 and the number 8 in the group to the left becomes a 9.

123 x 321

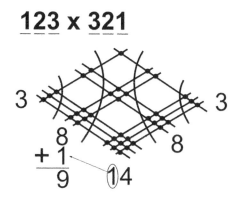

Now we can determine the answer by combining the numbers in each group. We write 123 × 321 = 39,483. And once again, that's the correct answer!

123 x 321 = 39,483

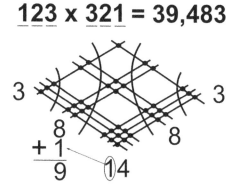

This chapter introduced multiplying by lines with a few examples. At this point, you probably have a sense of how to multiply by lines and might want to try out some problems. There is something fun about drawing lines and counting intersections to do multiplication.

Don't worry if you make mistakes at this point. Part of learning this method, or doing math in general, is to experiment and discover what works.

We suggest you try a few of the exercises. If you prefer, you can read the next chapter which describes the method in more detail and then return to the problems. The answers to these exercises appear in the last chapter.

Exercises

13 × 21

21 × 31

31 × 12

12 × 14

14 × 21

21 × 41

41 × 12

121 × 112

111 × 112

211 × 121

Chapter 2: The Steps To Multiply By Lines

Multiplying by lines is a process of drawing lines, grouping intersections between the lines, and then counting the intersections in each group. In order for this method to work, you must correctly draw the lines, make the proper groupings, and carry over numbers greater than 10 when counting the intersections.

We will explain the steps in detail, first by explaining multiplication of numbers with only one digit and then extending to numbers with two and three digits.

2.1 Single Digit Numbers

This is the easiest case. Let's say you want to multiply 3 by 8. You should draw 3 lines in one direction and then draw 8 lines in the other direction.

3 x 8 = 24

All of the intersections belong to a single group. There are 24 intersections and indeed $3 \times 8 = 24$.

Multiplying by lines works for single digit numbers. However, I personally do not recommend the method. I think it is very, very important to memorize the times tables up to 10. Knowing the times table is a useful skill and it is also helpful when multiplying by lines with larger digits, as we will explain in a later chapter.

While you can multiply single digit numbers by drawing lines, I recommend you do not and instead memorize the times table up to 10.

2.2 Two Digit Numbers

Let's say we want to multiply two numbers, AB times CD, where A, B, C and D are whole numbers between 1 and 9. There are 8 steps to multiplying by lines.

In all of the steps, try to draw all of the lines the same length.

1. Draw A lines slanted upward to the right.

In the drawings in this book, the lines are drawn at a 45 degree angle, parallel to each other, and equally spaced.

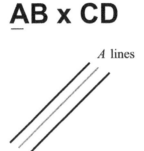

2. Next leave some space and draw B lines.

Try to draw the lines at the same angle so they are parallel to the A lines already drawn. How much space should you leave? In figures in this book, the bottom line of the B lines group is roughly aligned vertically with the top line of the group of A lines.

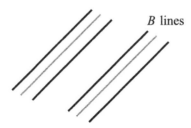

3. Now draw the lines for the number C in the opposite direction.

Draw C lines slanted downward to the right. Start drawing at the bottom left of the lines already drawn. In the figures in this book, the lines for the second number are drawn at a -30 degree angle.

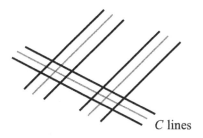

4. Finally, leave some space and draw D lines.

Try to draw the lines at the same angle so they are parallel to the C lines already drawn. The start of the D lines is roughly aligned vertically so as to make a diamond-like figure.

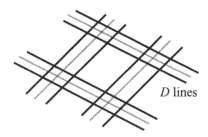

5. Make groupings of vertical intersections.

For AB times CD, there are three groups of intersections that are vertically aligned. There is a left group, a middle group consisting of two sets of intersections, and a right group.

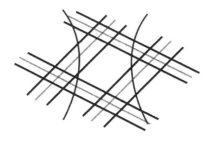

You are more than halfway done with the steps to multiply by lines! Now you need to interpret the figure to get to the answer.

6. Count the number of intersections in each group and write them down.

As this step depends on the particular numbers of the problem, we have not sketched an image. We suggest writing the number of intersections in each group in the figure, as we have been doing in the examples so far.

7. Carry over numbers if necessary.

This is the longest step to explain. You probably already know how to carry over numbers from the standard multiplication method. When multiplying by lines, the carry over process is the same. For completeness the steps are described here.

Carrying over is necessary only when the number of intersections in a group is 10 or greater. If every group has a number of intersections that is 9 or fewer, you can combine the numbers—from left to right—as your answer.

If any number is 10 or greater, however, then you will have to carry over the excess amount to the next group on the left.

Start with the group of intersections that is farthest on the right. If that number is 18, for example, you will cross that out and write 8. You should carry over the 1, which means to add the number 1 to the number in the middle group. In general, the number in the farthest right group will be between 1 and 81. If the number is between 1 and 9, you do not need to carry over. If it is between 10 and 19, carry over a 1; if it is between 20 and 29, carry over a 2; if it is between 30 and 39, carry over a 3; and so on, always carrying the value of the tens digit.

After you carry over to the middle group, if necessary, look at the resulting number in the middle group. Again, if the number is 10 or larger, you will carry over the value of the excess digits. In the middle group it is possible to get three digit numbers like 123. In that case, you will need to carry over the "12" to the left group. That is, cross out the number 123 and just write 3. Then you will add 12 to the number for the group to its left.

The number in the left group, after any carry over amount, is left alone. If the number is larger than 10 then just write that in your answer.

8. Write the answer from left to right.

The answer to AB times CD is written based on the numbers after the carry over process. Write the numbers from left to right as a single answer.

Let's do an example to illustrate carrying over numbers across multiple groups. Let's do 13 × 16.

You can sketch out the figure. We will describe the process in detail using the steps just explained.

First we draw the number 13 by drawing 1 line, leaving some space, and then drawing 3 lines. From steps 1 and 2, we draw the lines at a 45 degree angle, with the bottom of the 3 lines roughly aligned vertically with the end of the 1 line, and all of the lines are parallel to each other. From steps 3 and 4, we draw the number 16. We draw these lines at a -30 degree angle and leave space between the 1 line and the 6 lines.

Now we group the intersections that are vertically aligned, as described in step 5. There are groups of intersections on the right, in the middle, and on the left. Then step 6 is to count the number of intersections in each group. There are 18 intersections on the right, then 9 in the middle, and 1 on the left. At this point, the figure should look something like the following.

13 x 16

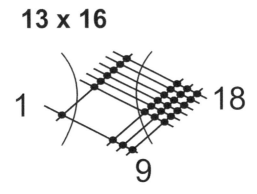

Now we need to deal with the carry over process described in step 7. We start with the number on the right. The number 18 is larger than 9 so we need to carry over the tens digit 1. We cross out the number 18 and write 8. We will add the 1 to the number in the middle group.

13 x 16

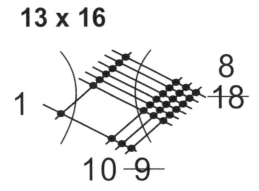

Now the number in the middle group is 10, which is larger than 9. Therefore, this number too will have to be adjusted by the carry over process. We cross out the number 10 and write 0. We will need to add the 1 to the number in the left group.

13 x 16

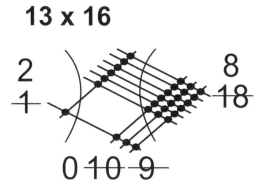

Now we are ready to write the answer. We combine the numbers from the left, the middle, and the right—in that order—to make a single number.

13 x 16 = 208

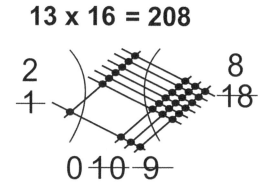

Therefore we have 13 × 16 = 208.

As this example illustrates, when you carry over numbers from multiple groups, it is important to organize the numbers and write them legibly. In the carry over process, a mistake in the early steps will "carry over" to the remaining steps.

2.3 Three Digit Numbers

With three digit numbers, the process is similar. The only part that requires special attention is how you draw the lines and make the groups of intersections. Rather than explaining with an abstract example of a problem like ABC × DEF, we will do the concrete example of 123 × 321.

Let's say you want to draw the number 123. We first draw 1 line for the 1.

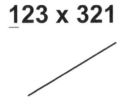

Now we need to leave some space and draw 2 lines for the number 2. Where should we draw these lines?

In our *two* digit example, we thought about each line as having *two* parts: a start and an end. So we aligned the start of second set of lines roughly with the end of first set of lines (in 13 we drew the 3 lines so the start of the last line was roughly vertically aligned with the end of the 1 line).

In a *three* digit number, we need to think about each line as having *three* parts: a start, a midpoint, and an end.

When we draw lines for each subsequent digit, we need to shift over by one part. So the second digit should have its last line start roughly at the midpoint of the first digit's set of lines. Drawn correctly, this will also make the midpoint of the second set of lines roughly align with the end of the first digit's lines. The drawing only has to be roughly correct, as illustrated in the following figure which is not perfectly aligned.

##

The final digit is shifted over by about the same distance. That means the bottom line in the group is roughly aligned vertically with the midpoint of the second set of lines and the end of the first set of lines.

##

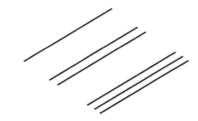

We draw the lines for the other number in a similar process. For the number 321, the lines for 3 start out on the far left and at the bottom, sloping downward at a -30 degree angle.

123 x 321

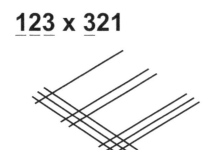

The lines for the digit 2 should be in the middle and shifted over so that their starting point is roughly the midpoint of the 3 lines just drawn, and their midpoint is roughly the ending of the 3 lines just drawn.

123 x 321

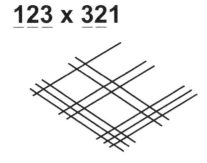

The final digit 1 will be drawn as 1 line and again shifted over, so that its far left is roughly aligned vertically with the midpoint of the 2 lines just drawn, and its midpoint is roughly aligned with the far right of the 2 lines.

123 x 321

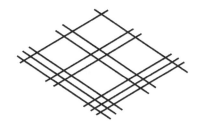

Drawn correctly, we can identify the groups of intersections that are vertically aligned. There will be 5 groups in total: a far left, a middle left, a middle, a middle right, and a far right.

123 x 321

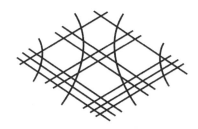

When multiplying two numbers that have 3 digits each, there is a lot more room to make mistakes when drawing the figure. Make the sketch carefully and check your work while making the groups.

Now we count the intersections in each group. Starting from the far right, there are 3, 8, 14, 8, and finally 3.

123 x 321

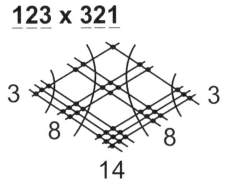

The carry over process is the same as in the two digit case. Start with the far right group of intersections and carry over to the group to its immediate left. Then continue to the next group and keep carrying over any results greater than 10 until the final group on the far left. The number in the group farthest left is written as is, no carry over is necessary.

123 x 321

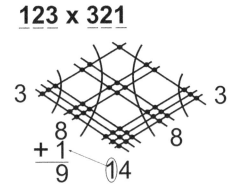

The final answer is determined just like in the two-digit example by combining the numbers after the carry over process, from left to right.

The numbers are 3, 9, 4, 8, and 3, which makes for 39,483.

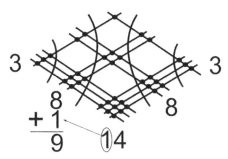

Again we will emphasize it is important to draw the figure carefully and write the numbers legibly during the carry over process.

2.4 More Than 3 Digits

If you're multiplying more than 3 digit numbers, it can get cumbersome to draw the figure.

However it is possible and the process is the same. The only important part is to keep the lines aligned correctly.

In general, if there is a number with N digits, you will want to divide the line mentally into N equally spaced division points running from start to end. After the first line is drawn, each subsequent line is shifted over by roughly one spot more in the N equally spaced division points.

When drawn correctly, you can group the intersections that are vertically aligned. When both numbers have two digits, there are 3 groups. When both numbers have three digits, there are 5 groups. When both numbers have four digits, there are 7 groups. In general, when both numbers have N digits, there are $2N - 1$ groups.

Here is an example of 1234×4321. Notice how each of the lines is roughly shifted over and how there are 7 vertically aligned groups.

1234 x 4321

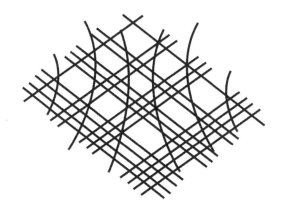

You can count out the intersections in each group and carry over to determine the answer for this problem.

As this example illustrates, multiplying by lines still works, but it gets harder to draw the figure for larger numbers.

Exercises

13 × 42

66 × 11

14 × 23

28 × 22

75 × 31

412 × 212

112 × 333

123 × 123

1111 × 2222

1212 × 2121

Chapter 3: Special Cases

Until now almost all of the examples have involved numbers without the digit 0, numbers that have an equal number of digits, and numbers whose digits are between 1 and 5. We deal with these cases now.

3.1 The Digit 0

Let's start off with an example of a number with a 0 digit. Let's do the problem of multiplying 13 by 30. We start out by drawing the lines for 13 as usual, drawing 1 line and then leaving some space to draw 3 lines.

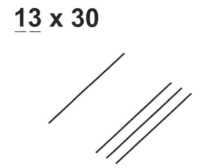

Then we draw the 3 lines for the "3" in 30 in the opposite direction.

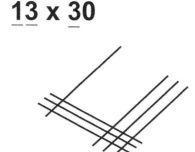

Now we have to draw a line for the "0" in 30. How do we do that? The trick we will use is to draw a special line for 0. We will use a dashed line to indicate a 0 line.

13 x 30

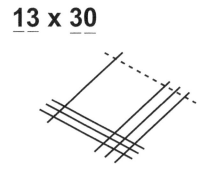

Now we can make groupings of vertically aligned intersections.

13 x 30

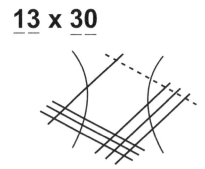

We are ready to count the intersections in each group. But now we need a special rule. With the 0-line, the rule is that we do not count any intersections with the 0-line. The logic is 0 times any number is 0, so any time the 0-line crosses another line, the number of intersections should be counted as 0.

Let's count the number of intersections. On the right there are three lines that intersect with the 0-line. The rule is that we do not count any of these intersections, so the number of intersections on the right is counted as 0.

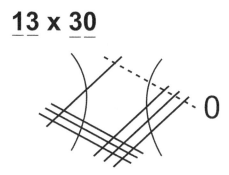

In the middle there are two groups of intersections. On the bottom, there are 9 intersections. On the top, the intersection involves the 0-line so we do not count it. Therefore, there are 9 intersections in the middle.

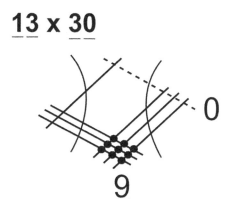

Finally, on the left there are 3 intersections.

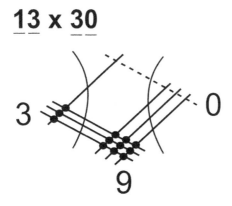

Now we can get to our answer as usual by combining the numbers of 3, 9, and 0. So we have 13 × 30 = 390.

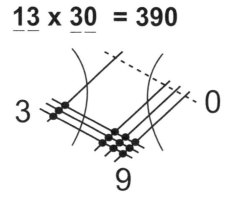

You might be wondering why we bother to draw the 0-line if we do not count its intersections. The reason is that multiplying by lines depends on drawing the figure correctly. You will want to draw a dashed line for the 0-line so that all of the lines get positioned correctly. We will do another example to illustrate this point.

Let's do the problem of 123 × 102. We start out by drawing the lines for 123.

123 x 102

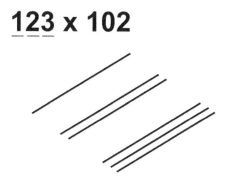

Now we draw 1 line for the "1" in 102.

123 x 102

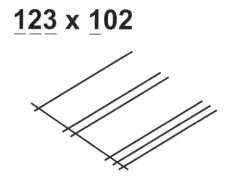

So far everything we have done is routine. The key step is the next digit in 102 which is the digit 0. If you simply omit this line, you might get in trouble when deciding where to draw the lines for the digit 2. So we will go ahead and draw the digit 0 using a dashed line.

The dashed line for the digit 0 is placed exactly where we would draw the lines for the second digit of the second number.

123 x 102

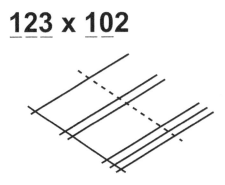

Finally we draw 2 lines for the "2" in 102.

123 x 102

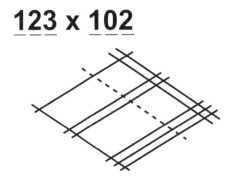

The figure drawn is very similar to the one we drew for 123×321. So we can finish solving this problem just like we did when all of the numbers were positive.

We will mark out the five groups of vertically aligned intersections.

123 x 102

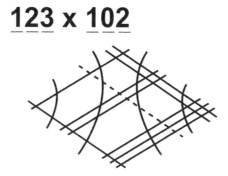

Finally we count intersections, except we do not count the intersections made with the 0-line.

Starting from the right, there are 6 intersections, 4, 5, 2, and finally 1. We will go ahead and write these intersections in the figure. We get to the answer by combining the numbers from left to right. So our answer is 12,546.

123 x 102 = 12,546

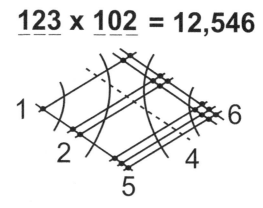

In this example, you can see the importance of drawing a dashed line for the 0. That dashed line was a placeholder so the rest of the lines could be aligned correctly.

3.2 Three Digits Times Two Digits

Now we move on to another topic. So far we have been multiplying numbers that have the same number of digits. But what happens when we want to multiply 213 by 12? Now we have a three digit number multiplied by a two digit number. How we do we do that?

Here is the trick. We will re-write 12 as a three digit number 012 that includes the leading digit 0. Now we can solve the problem of 213×012 because we already know how to multiply two 3-digit numbers and we just explained how to multiply numbers that involve the digit 0.

Let's solve this problem. Start by drawing the lines for 213.

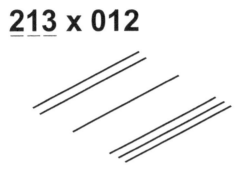

Now we need to draw 012. First draw a dashed line for the "0" in 012.

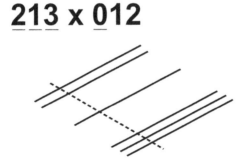

Then draw 1 line for the "1" in 012.

2̲1̲3̲ x 0̲1̲2̲

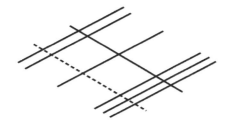

Finally draw 2 lines for the "2" in 012.

2̲1̲3̲ x 0̲1̲2̲

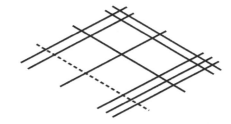

Then make groupings of vertically aligned intersections.

2̲1̲3̲ x 0̲1̲2̲

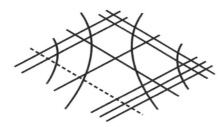

Now count the number of intersections, remembering not to count any intersections made with the 0-line. Starting from the far right, there are 6 intersections, 5, 5, 2, and finally 0. The digits are 02556. Typically we do not write leading 0's in a number.

So we can write the answer as 213 × 012 = 2,556.

213 x 012 = 2,556

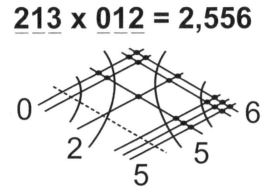

The example shows one way to solve the problem. There is a shortcut to this method. You could remember the 2-digit number should get aligned to the right side. So after practice you might be able to avoid drawing the 0-line and just draw the following figure for 213 × 12.

213 x 12 = 2,556

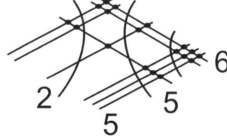

How you solve the problem is up to you.

However, when learning how to multiply a 3-digit number by a 2-digit number, it is generally good practice to write the 2-digit number with a leading 0 and draw the 0-line so the groups are aligned correctly.

3.3 A Large Digit

Let's say you want to multiply 12 by 19. Start by drawing the lines for the number 12.

12 x 19

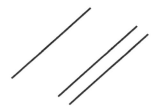

Then draw 1 line for the "1" in 19.

12 x 19

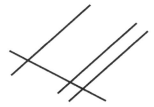

At this point, normally we would draw 9 lines for the "9" in 19. But if we do that, the figure would get congested and there would be many intersections to count.

A trick for large digits is to make a special line for each large digit. Just like we used a special dashed line for the digit 0, we can use a special thick line for a digit that is large. We also need a special rule to count intersections: any time the 9-line crosses a thin line, it should count for 9 intersections.

So let's try this out.

We will draw a single thick line for the digit 9.

12 x 19

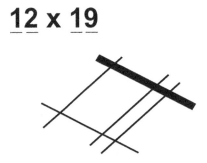

So that we remember that each time the thick line crosses the thin line it should count as 9 intersections, we will write in a small 9 in the figure next to each intersection of the thick line and the thin lines.

12 x 19

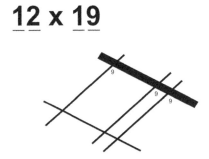

This extra annotation will help us when we count the number of intersections.

Now let's make the groupings of vertically aligned intersections.

12 x 19

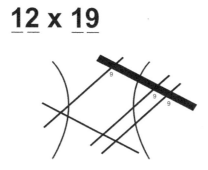

Finally, we count the number of intersections. On the right, there are two intersections of the 9 line. Each intersection is equivalent to 9 intersections, so we add up 9 + 9 = 18. There are 18 intersections in total.

In the middle, the top is the intersection of a 9-line with a single line, so that counts for 9 intersections. On the bottom there are two intersections of thin lines, so they contribute 2 more intersections. In total there are 11 intersections.

And on the left there is 1 intersection.

12 x 19

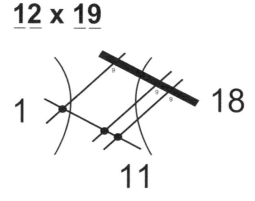

Now we carry over the numbers from each group. The group on the right has 18, which means we need to carry over the 1 and leave an 8. The next group becomes 12.

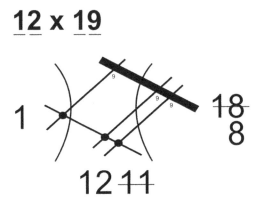

The group in the middle has a 12, so we need to carry over the 1 to the group on the left and leave a 2.

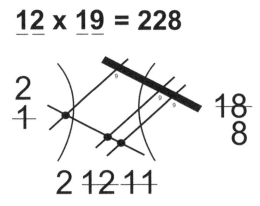

The digits in each group of numbers are 2, 2, and 8. So we combine these numbers from left to right to get the answer of 228.

Using a thick line for a large digit is a trade-off. We get the advantage of drawing fewer lines at the cost of making the counting a bit harder. Still, since it keeps the figure less cluttered, it is generally worth doing.

3.4 Many Large Digits

Let's say you wanted to multiply 97 and 96. You can definitely do this problem by drawing the number 97 as 9 lines and then 7 lines, and the number 96 as 9 lines and then 6 lines. But as the last section illustrates, it can be useful to treat each larger digit specially with a thick line.

When we use thick lines, we will run into cases where two thick lines intersect each other. How do we treat a 9-line intersecting with another 9-line? Or a 9-line intersecting with a 7-line?

Remember that each 9-line represents 9 individual lines. So when two 9-lines intersect each other, it is the same as if we drew out 9 individual lines and intersected that with another 9 individual lines. How many intersections are there? Well, drawing 9 individual lines by 9 individual lines is exactly the diagram we would draw for the problem 9 times 9. Therefore, there are $9 \times 9 = 81$ intersections.

This leads us to a general rule for dealing with larger digits. If we have the intersection of a thick line representing one number A, and another thick line representing B, then the number of intersections between those two lines is $A \times B$.

So let's return to the original problem of 97×96. Instead of drawing 9 lines and then 7 lines for 97, let's use the principle of drawing thick lines for larger digits. Draw a thick line for the number 9 and another thick line for the number 7. In this figure, we will mark the lines to remember the digits they represent.

97 x 96

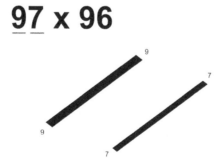

For the number 96, draw a thick 9-line and a thick 6-line.

97 x 96

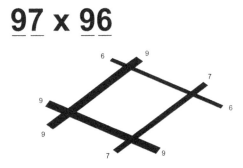

Now make the groupings of vertically aligned intersections.

97 x 96

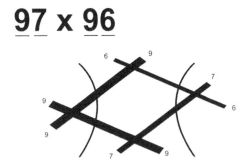

It is worth taking a moment to consider this figure. Because we are using thick lines, the figure looks very similar to the one we drew for our introductory example of 12 × 13. Instead of drawing 16 lines for the number 97 and 15 lines for the number 96, we could depict those 31 lines using a mere 4 lines. The trade-off is that now we have to count the intersections carefully instead of simply having each intersection count for 1 when we were using thin lines.

So let's count the number of intersections. On the right there is a 6-line intersecting with a 7-line. Whenever two thick lines intersect, the total number of intersections is equal to the product of those two numbers. So this counts as 6 × 7 = 42 intersections.

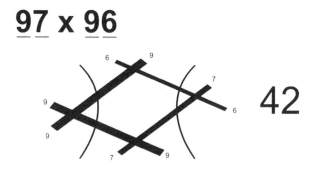

In the middle, the top has the intersection of a 9-line with a 6-line. This counts as 9 × 6 = 54 intersections. The bottom is the intersection of a 9-line with a 7-line, which counts as 9 × 7 = 63 intersections. So in all there are 54 + 63 = 117 intersections.

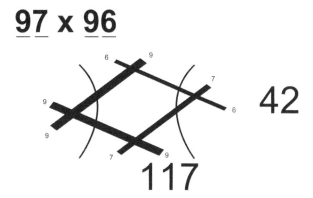

On the left is the intersection of two 9-lines so there are $9 \times 9 = 81$ intersections.

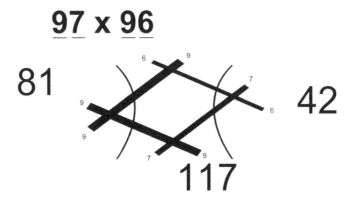

Now we need to carry over. Every group has a number that is larger than a single digit, so we have our work cut out ahead of us!

We start with the group on the far right and cross out the number 42. We will carry over the tens digit 4 to the next group and leave the digit 2.

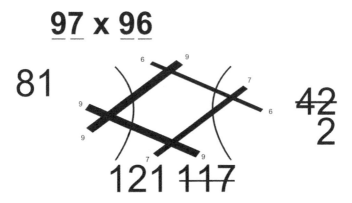

Already the diagram has a lot of numbers written, and we still have another carry over in the middle group for 121.

Now we cross out 121, and carry over the 12, leaving only the digit 1 in the middle group. (Remember we always carry over any result larger than 9. In this case we have to carry over two digits 12).

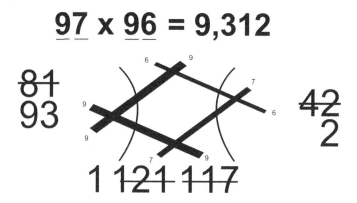

On the left the number is 93, in the middle 1, and on the right 2. We will combine those numbers to get to the answer of 97 × 96 = 9,312.

This problem illustrates several lessons. First, it is really important to know your times table up to 10 because you will need to know it to calculate the number of intersections between two large digits. Second, some problems involve carrying over many of digits. This means it is very important to draw the diagram neatly and take your time step by step since any early mistake will cause the final result to be incorrect.

Also, problems with larger digits tend to be harder to solve in general. You will need to know your times tables, and you will have to carry over many digits. This is equally true when multiplying using the standard method—in either case you will have to carry over digits so the problems tend to be harder.

With all of these difficulties, you might be better off multiplying by the standard method.

However, this example was done to illustrate it is possible to multiply numbers with large digits.

When should you use special lines for larger digits? There is no specific rule, but it is often helpful whenever the diagram might get cluttered. Very often this will be for digits that are larger than 5.

Exercises

40×31

201×312

203×102

321×53

52×59

78×93

69×32

89×97

47×28

19×18

Chapter 4: Rotating The Diagram

The first example in this book was 12×13. We drew out the following figure to multiply by lines.

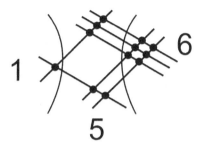

On the left there is 1 intersection, then in the middle there are $5 = 2 + 3$ intersections, and on the right there are 6 intersections. So the answer is $12 \times 13 = 156$.

While writing this book, I came across a fun discovery. I was sketching all of my diagrams on paper. I wondered, what would happen if I turned the paper and rotated the diagram? Would that lead to anything useful?

For example, here is what the figure for 12×13 looks like after being rotated 90 degrees clockwise.

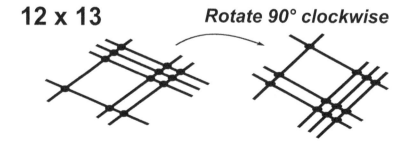

The new shape is slightly off-center because of the angles of the lines in the original figure. But looking a bit abstractly, the rotated figure resembles the original diagram. And that's the realization: when the figure for 12 × 13 is rotated, it creates a new figure that can solve a new problem!

But what does this new diagram mean? In all of the problems so far, we have been taking a problem like 12 × 13 and drawing lines for each of the digits. With the rotated diagram, we have already have drawn the lines. So we need to work backwards: instead of turning numbers into lines, we need to translate the lines in the diagram into numbers.

In the rotated diagram, look at the lines that are angled upward in the northeast direction. There is 1 line on the left, and after some space, there are 3 lines. We can read those lines as representing the number 13.

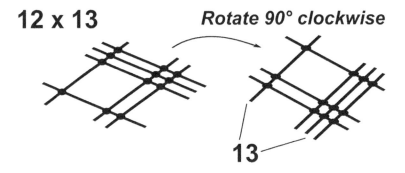

For the lines angled downward to the southeast direction, there are 2 lines on the bottom left, and after some space, there is 1 line to the top right. We can read those lines as representing the number 21.

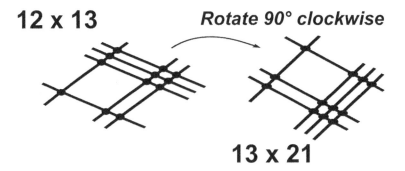

Therefore, we conclude this rotated figure corresponds to the problem of 13 × 21. Now comes the fun part. We can count the intersections in this rotated figure to solve this entirely new problem of 13 × 21!

Let's count the intersections. There are 2 intersections on the left, then there are 1 + 6 = 7 in the middle group (counting the top and the bottom), and finally 3 on the right. Combining the numbers as usual, this leads to the conclusion that 13 × 21 = 273. And that's the right answer!

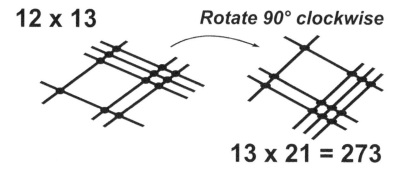

Like magic we have used the figure of 12 × 13 to solve an entirely different problem of 13 × 21.

But why stop there? Let's keep rotating! If we rotate the figure another 90 degrees clockwise we end up with the following figure.

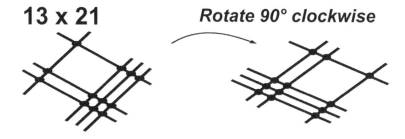

Once again we will convert the lines in diagram into a numerical problem. The lines sloping upward to the northeast are 2 lines on the left, and after some space, 1 line. These lines represent the number 21.

For the lines sloping downward in the southeast direction, there are 3 lines on the bottom left and after some space 1 line on the top right. So the other number is 31. Therefore, we conclude this diagram corresponds to 21 × 31.

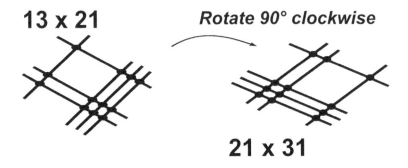

To solve the numerical problem, we count the intersections. On the left there are 6 intersections, in the middle there are 2 + 3 = 5 intersections (remembering the middle group has a top and a bottom), and on the right there is 1 intersection. Combining these numbers into a single answer, we conclude 21 × 31 = 651.

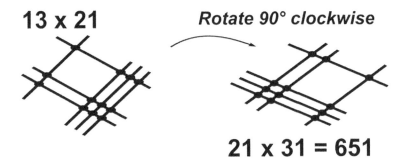

To recap, we have taken the original figure for 12 × 13 = 156 and derived answers for two entirely new problems. Namely, we have solved the problem 13 × 21 = 273 and another problem of 21 × 31 = 651.

There's no need to stop now! We can rotate the diagram one more time 90 degrees clockwise to create one more figure.

Here is what the diagram for 21 × 31 looks like when rotated 90 degrees clockwise.

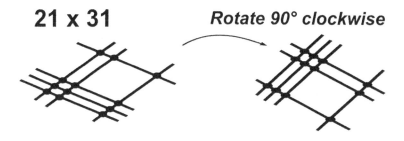

We can read the new figure as we have been doing.

You should see the pattern by now. The first number can be read by the lines sloping upward to the northeast. There are 3 lines followed by 1 line, which means the first number is 31. The lines in the other direction determine the digits for the other number. There is 1 line followed by 2 lines, so the other number is 12.

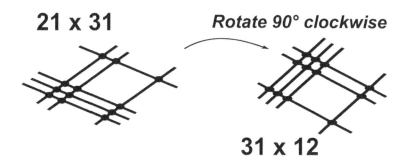

Rotating 21 × 31 thus leads to the problem of 31 × 12.

Now we count the intersections. There are 3 intersections on the left, a total of 6 + 1 = 7 in the middle, and 2 on the right. Therefore we have that 31 × 12 = 372.

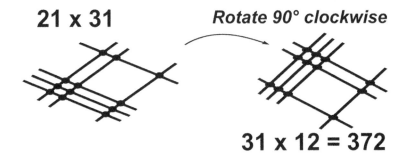

If we rotate the diagram one more time, we will return to the original figure for the original problem of 12 × 13.

Amazingly, the diagram for 12 × 13 allowed us to solve the related problems of 13 × 21, 21 × 31, and 31 × 12.

There is a concise way to summarize the result by writing all four problems on the same figure.

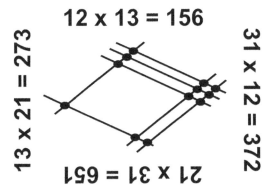

This diagram expresses all four of the problems the single figure can solve, with the problem solved written at the top (north side) for a particular orientation of the figure.

For a general problem of AB × CD, the figure rotated 90 degrees clockwise corresponds to CD × BA. Rotating another 90 degrees corresponds to BA × DC, and a third 90 degree rotation corresponds to DC × AB.

In other words, a 90 degree rotation clockwise switches the order of the digits in the first number, and then it flips the order of the two numbers being multiplied.

When A and B are equal, or C and D are equal, some of the rotations will not solve new problems. For example, the figure for 11 × 12 rotated 90 degrees clockwise results in the figure for 12 × 11. But since we have that 11 × 12 = 12 × 11, rotating the diagram does not solve a unique multiplication problem.

This chapter explained rotating a diagram for two digits numbers. You can also rotate the diagram when the problem involves three digit numbers. A 90 degree rotation clockwise switches the order of the first and last digits in the first number, and then it flips the order of the two numbers being multiplied.

Exercises

Solve each problem and rotate the diagram to solve three additional problems.

14 × 23

123 × 221

Chapter 5: Why Multiplying By Lines Works

We started the book with an example of 12 × 13. We drew 1 line for the "1" in 12, then we left some space and drew 2 lines for the "2" in 12. We did the same procedure for 13 by drawing the lines in the other direction. After we grouped vertically aligned intersections, we could come up with the answer.

12 x 13 = 156

So we found 12 × 13 = 156.

We have illustrated many problems in this book. The question is, why does multiplying by lines work? What is the reason we can draw lines, count intersections, and then produce an answer that's mathematically correct?

In this chapter we explain why the method works. We will start out with an example of the problem of 2 × 3 and then build up to the figure for the problem 12 × 13, explaining exactly what each line, the groups of intersections, and the number of intersections represent.

The abstract idea is that multiplying by lines is a shorthand for counting hundreds and thousands of intersections really quickly, and counting intersections is closely connected to multiplication. Let's get into the details.

5.1 Line Intersection Is Multiplication

Let's start out with the simple example of 2 × 3. We draw 2 lines and then 3 lines in the other direction.

If you count, there are 6 intersections, and in fact 2 × 3 = 6.

You can similarly do the same for any two numbers A and B, and you will always find that you will get A × B intersections.

Why is that?

Think about what happens when you draw A lines in one direction and B lines in another. Each of the lines for the number B crosses over a total of A lines. That means the first line crosses over A lines, the second line crosses over A lines, and so on. The total number of intersections is found by adding A intersections repeated B times. In other words, the total number of intersections is A + A + ... + A = A × B.

Therefore, drawing A lines and then drawing B lines in another direction is one way to do multiplication of whole numbers.

Now we need to explain how this relates to the figures we have been drawing.

5.2 Bundles Of Lines

Let's say you want to multiply 2 × 13. We could draw 2 lines in one directly and then draw 13 lines in the other direction.

2 x 13 = 26

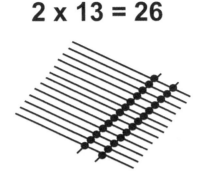

In total there are 26 intersections, and in fact $2 \times 13 = 26$.

Now we will simplify the figure. In the group of 13 lines, let's separate 10 of them and bundle them together. Think about each line like a stick, and we will be taking 10 of the sticks and wrapping them in a bundle.

We draw this bundle of 10 lines as a thick line.

2 x 13 = 26

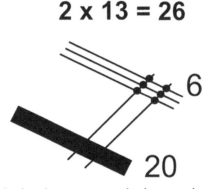

The diagram also depicts how to count the intersections in this figure. On the upper right we have thin lines that contribute 6 intersections. The bundle, which represents 10 lines, intersects the 2 thin lines as well. Since the bundle is equal to 10 lines, each time the bundle crosses a thin line represents 10 intersections. Therefore, there are 20 intersections between the bundle of 10 lines and the 2 thin lines. We can add up the 20 intersections and the 6 intersections to see there are 26 intersections in all. So this figure also shows that $2 \times 13 = 26$.

Now let's add another twist. Instead of the multiplying 2 by 13, let's multiply 12 by 13.

Instead of starting over, we can modify the figure for 2 × 13. We could draw an extra 10 lines parallel to the 2 lines already there. But what we will do instead is add a single bundle of 10 lines.

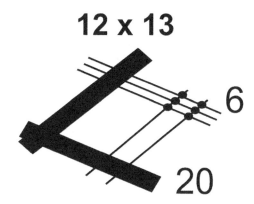

Notice our figure is starting to resemble the method of multiplying by lines!

The number 12 is drawn as a thick line, representing 10 lines, followed by some space and 2 thin lines. The number 13 is drawn in the other direction, depicted as a thick line for 10 lines and after some space 3 thin lines.

Do you see where we are going? We will now need to count the number of intersections of the thick line for the number 12 with the existing lines.

In the middle top, the thick line intersects with 3 lines. Since the thick line represents 10 lines, each time the thick line crosses the thin line that counts for 10 intersections each. So we have an extra 30 intersections. We will write that in the figure.

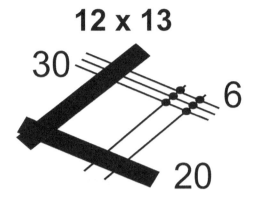

Finally we have the intersection of two thick lines on the left. We know each thick line is 10 lines, so we really have the intersection of 10 lines with another 10 lines. This means there are 10 × 10 = 100 intersections for the two thick lines, which we write in the figure.

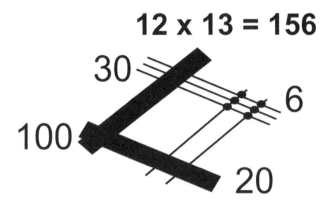

How many intersections are there for 12 × 13? We can add up the numbers to count the total number of intersections. We can count that there are 100 + 50 + 6 = 156 intersections, and that is 12 × 13.

We could multiply by lines using the above technique—drawing thick bundled lines, counting the number of intersections, and adding them up. After doing this method over and over, there are simplifications we can make.

First, we can drop the 0 when counting the intersections in the middle. The middle intersections will always be a group that has thick 10-lines and thin single lines, contributing intersections in multiples of 10. All multiples of 10 have a 0 in the tens group, so we can drop the 0 and just remember the result is a multiple of 10.

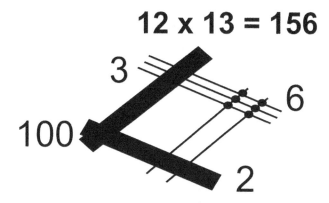

Similarly we can drop the 00 on the left. The group on the left always has the intersection of thick lines. When two thick bundled 10-lines intersect, they will contribute intersections in multiples of $10 \times 10 = 100$ intersections. Therefore, the we can omit the 00 in the number and remember the result is a multiple of 100.

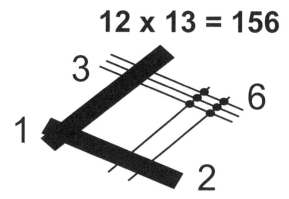

Now we make another adjustment. We have drawn the thick lines to represent bundles of 10 lines. We can also draw them as thin lines, and remember they represent 10 lines that we have separated from the other thin lines. The spacing of the lines allows us to avoid having to draw thick lines.

12 x 13 = 156

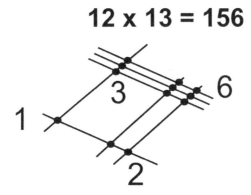

Finally, we can group the vertically aligned intersections. Now it is clear what these vertically aligned groups represent. The group on the right represents intersections of lines for multiples of 1, the group in the middle represents multiples of 10, and the group on the left represents multiples of 100.

12 x 13 = 156

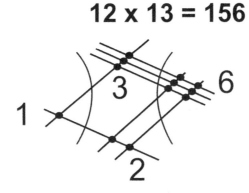

And now we see where the figure for 12 × 13 comes from and what it means. We have drawn 12 lines and then 13 lines using shorthand. We can determine the answer by counting the groups of intersections: on the right is the units place, in the middle is the tens place, and on the left is the hundred's place. This also explains the carry over rule: since each group is a placeholder for a power of 10 (the units digit, the tens digit, the hundreds digit, etc.), the maximum value of a particular group (except the far left one) has to be 9. Any excess digits have to get carried over to the group on the left.

When we draw the figure for 12×13, we are really counting 156 intersections by the shorthand of counting 6 intersections, then 50 intersections, and then 100 intersections. Because the placeholders are always the same, we just write 6 on the right, 5 in the middle, and 1 on the left.

When multiplying 3-digit numbers, multiplying by lines works for a similar reason. For the number 123, we could draw 123 individual lines. Instead, we draw a single line for the 1 to mean a bundle of 100 lines and then two lines for 2 to mean two bundles of 10 lines each. Finally we draw 3 lines that are each single lines for the number 3.

In summary, multiplying by lines is an artistic shorthand to count hundreds and thousands of intersections very quickly, and counting intersections is the same process as multiplying two whole numbers.

Chapter 6: Advanced Examples

In this chapter we explore some special applications of multiplying by lines. These are not always the most practical ways to solve the problems, but they build upon what we have learned so it is useful to mention them.

6.1 Algebraic Expressions

Multiplying by lines can also be used for algebraic expressions. Let's say you wanted to multiply $(x + 2y)(x + 3y)$. We first draw 1 line for the x in $(x + 2y)$.

$$(\underline{x} + 2y)(x + 3y)$$

To be clear, let's explain what that single line represents. The quantity x can also be written as $1x$. So we are drawing the single line for the coefficient of 1 in the value of $1x$, and if we had $2x$ we would need to draw 2 lines.

This will lead to a procedure for drawing the rest of the lines. We need to draw lines to match the value of the coefficient of the variable.

Continuing the process, we need to leave some space and draw 2 lines for the 2y in (x + 2y). We draw the lines for y as slightly thicker to distinguish them from the lines for x.

$(\underline{x} + \underline{2y})(x + 3y)$

For the number (x + 3y), we draw the lines in the opposite direction. Let's draw 1 line for the x.

$(\underline{x} + \underline{2y})(\underline{x} + 3y)$

Note that we draw this figure just as we have been drawing all of the other problems in this book.

Finally we draw 3 lines for the 3*y*.

(*x* + 2*y*)(*x* + 3*y*)

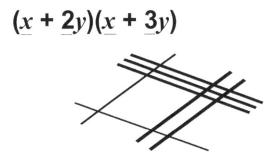

Now let's count the intersections. We will use the rule already established. Whenever two lines intersect, the result is the product of the values for the lines they represent.

For the group on the right, the intersection is between *y*-lines so the result will be a multiple of $y \times y = y^2$. Because there are 6 intersections, the result is $6y^2$.

(*x* + 2*y*)(*x* + 3*y*)

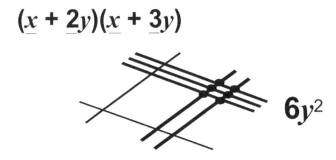

$6y^2$

In the middle, the intersections are between *x*-lines and *y*-lines. This means the result will be a multiple of $x \times y = xy$.

Since there are 5 intersections, the result is $5xy$.

$(x + 2y)(x + 3y)$

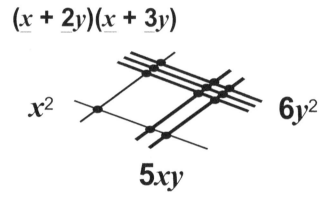

$5xy$

$6y^2$

On the left, the intersection is between two x-lines, so the result is x^2.

$(x + 2y)(x + 3y)$

x^2 $6y^2$

$5xy$

Notice that like other problems in the book we also have grouped up the vertically aligned intersections in the middle group. We have so far omitted the lines for the groupings because the thick and thin lines associate the intersections with particular variables. But we will draw the lines in the next figure to emphasize the groupings of variables.

How many intersections are there in total? Adding up these values gets us to the answer. We have $(x + 2y)(x + 3y) = x^2 + 5xy + 6y^2$.

$(x + 2y)(x + 3y) = x^2 + 5xy + 6y^2$

x^2 $6y^2$

$5xy$

When multiplying algebraic expressions, note there is no carrying over to the next group for values larger than 10 because each vertically aligned group of intersections is associated with a specific product of variables. So if you do $(x + 3y)(x + 4y)$, the group of intersections on the far right has $3 \times 4 = 12$ intersections, and the intersections are counted as $12y^2$.

6.2 Complex and Imaginary Numbers

For those familiar with complex numbers, you can also use multiplying by lines for the imaginary number i where $i^2 = -1$.

Let's say that we want to expand $(3 + i)(1 + 2i)$. We first draw 3 lines for the 3 in $(3 + i)$.

$(3 + i)(1 + 2i)$

Then we leave some space and draw 1 line for the *i* in (3 + *i*). We draw the lines for *i* as slightly thicker to distinguish them with the lines for real numbers.

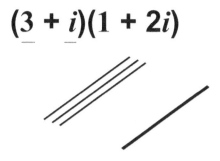

To be clear, we are drawing this single line for the coefficient of 1 in the expression for *i*, since $i = 1i$.

For the number (1 + 2*i*), we draw the lines in the opposite direction. Let's draw 1 line for the 1 in the expression.

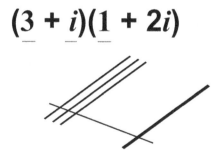

Then we draw 2 lines for the 2*i*.

(3 + *i*)(1 + 2*i*)

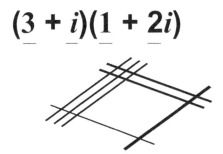

Now let's count the intersections. On the right, we have the intersection of two *i*-lines, so the result is a multiple of $i \times i = i^2 = -1$.

Since there are 2 intersections, the result is $2i^2 = 2(-1) = -2$.

(3 + *i*)(1 + 2*i*)

$2i^2 = -2$

Now oddly we have a negative number in one of our groupings. That will ultimately pose no problem. We will simply need to add the value of negative 2 to the results from counting intersections in the remainder of the problem.

In the middle, we have the intersections of an *i*-line with lines representing the number 1. So the result will be a multiple of $i \times 1 = i$.

Since there are 7 intersections in the middle group, the result is $7i$.

(3 + *i*)(1 + 2*i*)

$2i^2 = -2$

$7i$

On the far left, there are intersections between two lines representing the number 1. So the result is a multiple of $1 \times 1 = 1$.

Since there are 3 intersections, the result is 3.

(3 + *i*)(1 + 2*i*)

3

$2i^2 = -2$

$7i$

Adding up these values gets us to the answer.

We have the result $(3 + i)(1 + 2i) = 3 + 7i - 2 = 1 + 7i$.

$$(3 + i)(1 + 2i) = 1 + 7i$$

$$3 \qquad\qquad 2i^2 = -2$$

$$7i$$

This is the correct answer and the negative value of -2 simply subtracted from the positive number 3 for the group on the left.

6.3 Negative Numbers

Some algebraic expressions involve negative quantities. The rule for counting intersections is the same: whenever two lines cross, the result is the product of the two lines.

When two lines are both positive or both negative numbers, the intersection should be counted as a positive number. When one is positive and the other negative, the intersection is counted as a negative.

Let's say you wanted to multiply $(x - 2y)(x - 3y)$. The figure will look much like the one for $(x + 2y)(x + 3y)$. The only difference is we draw the negative lines as dashed lines, and we keep track of the signs of the results. When two negative lines cross they make a positive, when a positive line and a negative line cross it is a negative, and when two positive lines cross the result is a positive.

$$(x - 2y)(x - 3y) = x^2 - 5xy + 6y^2$$

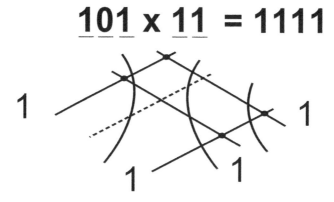

The above figure shows the result $(x - 2y)(x - 3y) = x^2 - 5xy + 6y^2$.

This example illustrates that you can extend multiplying by lines to include negative numbers like -1, -2, and so on.

6.4 Binary Arithmetic

This is an advanced topic if you are familiar with other number bases. Let's say that you wanted to multiply 101 by 11 in binary (base 2 arithmetic).

Let's draw 101 × 11 as usual, and make the groupings.

101 x 11 = 1111

There is exactly 1 intersection in each group, so the answer is 1111. This happens to be the correct answer for the problem, and multiplying by lines will work for other number bases besides the decimal system.

To multiply numbers in binary, the procedure is exactly the same. The only difference is that now you need to carry over digits whenever the number of intersections is greater than 1. So if a group has 3 intersections, you would write that in binary as 11 and carry over a 1 to the next group. Or if a group had 4 intersections which is 100 in binary, you would carry over the 10 and leave the digit 0 in that group.

The same process can work for other number bases. Draw the lines as you normally would, and count the intersections—but write them in the appropriate number base. Then carry over whenever the number of intersections is more than a single digit (in the appropriate number base). The final results, read from left to right, is your answer.

The exercises to this chapter only pertain to expanding a few algebraic expressions to illustrate the technique. The reason is that when expanding algebraic expressions, complex numbers, or doing binary arithmetic, the standard methods taught in schools are generally more efficient and we encourage you to learn those.

Exercises

$(2x + 3y)(x + y)$

$(x + y)(x + y)$

$(x + y)(x - y)$

Conclusion

Multiplying by lines is a wonderful artistic method to multiply two and three digit numbers. The method works because it is a shorthand for counting up hundreds and thousands of intersections between lines. On occasion, you will find it useful to draw thicker lines to represent digits larger than 5, and you will need to draw a dashed line for the digit 0.

Once you draw a figure of a numerical problem, you can rotate the figure and solve up to three additional multiplication problems. Multiplying by lines can also work for expanding algebraic expressions, multiplying complex numbers, and doing arithmetic in binary and other number bases.

Multiplying by lines is a magical and mysterious technique for solving problems. It is often difficult to get people interested in math. This is one topic that is exciting and I hope you enjoyed learning more about it.

74

Chapter 7: Answers To Exercises

Chapter 1

13 x 21 = 273

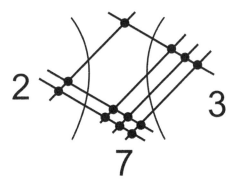

21 x 31 = 651

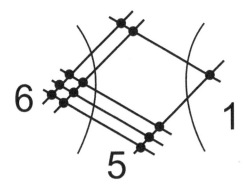

31 x 12 = 372

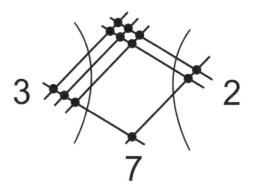

12 x 14 = 168

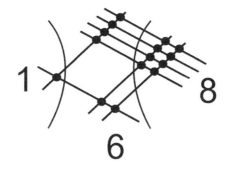

14 x 21 = 294

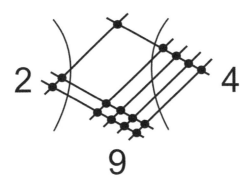

21 x 41 = 861

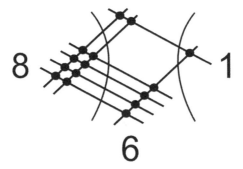

41 x 12 = 492

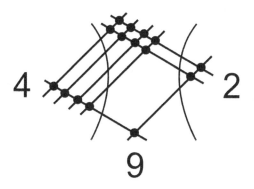

121 x 112 = 13,552

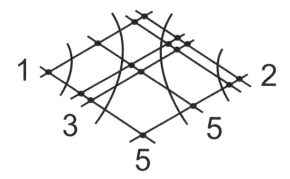

111 x 112 = 12,432

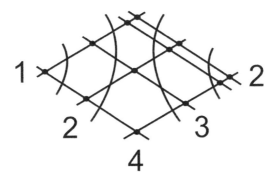

211 x 121 = 25,531

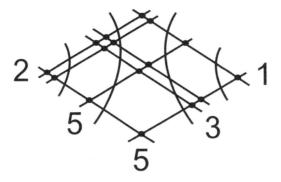

Chapter 2

13 x 42 = 546

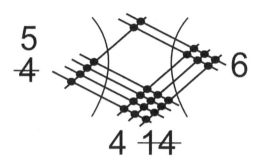

66 x 11 = 726

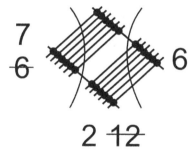

14 x 23 = 322

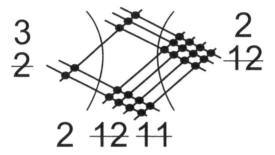

28 x 22 = 616

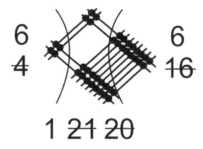

75 x 31 = 2,325

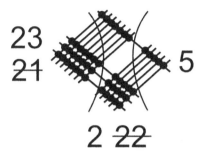

412 x 212 = 87,344

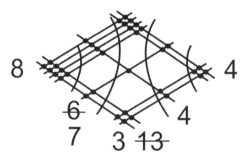

112 x 333 = 37,296

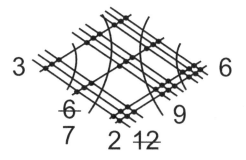

123 x 123 = 15,129

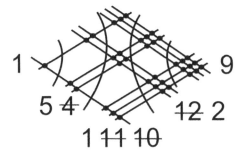

1111 x 2222 = 2,468,642

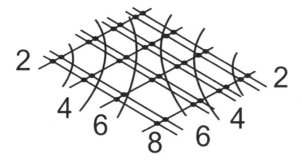

1212 x 2121 = 2,570,652

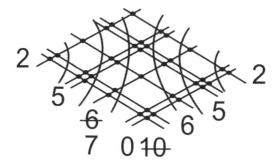

Chapter 3

40 x 31 = 1,240

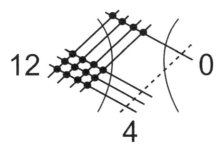

201 x 312 = 62,712

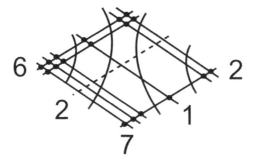

203 x 102 = 20,706

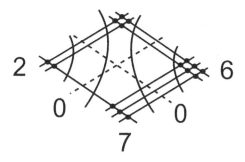

321 x 53 = 17,013

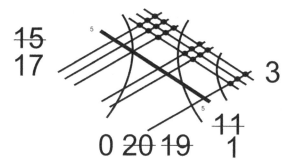

52 x 59 = 3,068

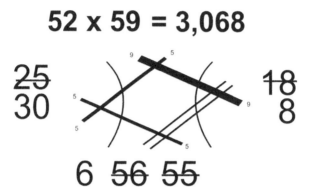

78 x 93 = 7,254

69 x 32 = 2,208

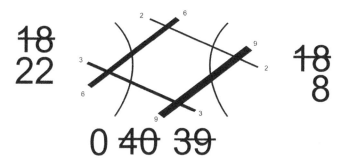

89 x 97 = 8,633

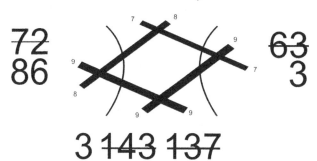

47 x 28 = 1,316

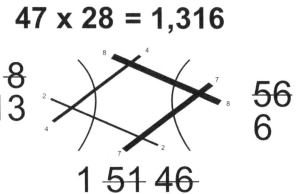

19 x 18 = 342

Chapter 4

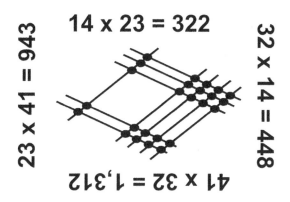

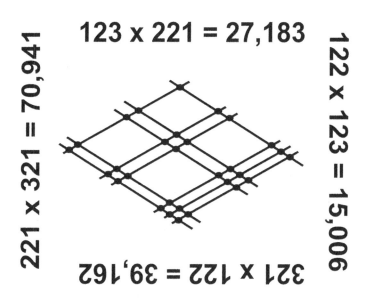

Chapter 6

$$(2x + 3y)(x + y) = 2x^2 + 5xy + 3y^2$$

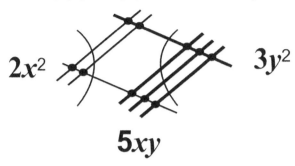

$$(x + y)(x + y) = x^2 + 2xy + y^2$$

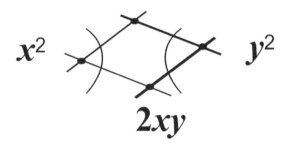

$$(x + y)(x - y) = x^2 - y^2$$

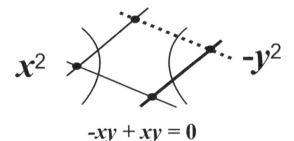

$$-xy + xy = 0$$

More from Presh Talwalkar

I hope you enjoyed this book. If you have a comment or suggestion, please email me presh@mindyourdecisions.com

Books by Presh Talwalkar

The Joy of Game Theory: An Introduction to Strategic Thinking. Game theory is the study of interactive decision-making, situations where the choice of each person influences the outcome for the group. This book is an innovative approach to game theory that explains strategic games and shows how you can make better decisions by changing the game.

Math Puzzles Volume 1. This book contains 70 interesting brain-teasers and fun math problems in counting, probability, and game theory.

But I only got the soup! This fun book discusses the mathematics of splitting the bill fairly.

40 Paradoxes in Logic, Probability, and Game Theory. Is it ever logically correct to ask "May I disturb you?" How can a football team be ranked 6th or worse in several polls, but end up as 5th overall when the polls are averaged? These are a few of the thought-provoking paradoxes covered in the book.

About The Author

Presh Talwalkar writes the *Mind Your Decisions* blog, a website with math puzzles and articles on game theory. The *Mind Your Decisions* YouTube channel features videos on math. He studied Economics and Mathematics at Stanford University.

Made in United States
Orlando, FL
04 September 2022